GETTING THE JOB DONE

PLUMBERS AND SEWAGE WORKERS

Nathan Miloszewski

PowerKiDS press

New York

Published in 2020 by The Rosen Publishing Group, Inc.
29 East 21st Street, New York, NY 10010

First Edition

Editor: Greg Roza
Book Design: Reann Nye

Photo Credits: Cover sturti/Getty Images; pp. 5, 7 Andrey_Popov/Shutterstock.com; pp. 8, 9, 16 Vladimir Mulder/Shutterstock.com; p. 10 micmacpics/Shutterstock.com; p. 11 Bertl123/Shutterstock.com; p. 13 Richard Levine/Corbis News/Getty Images; p. 15 sturti/E+/Getty Images; p. 17 Ethan Miller/Getty Images News/Getty Images; p. 18 BartCo/E+/Getty Images; pp. 19, 21 Monkey Business Images/Shutterstock.com; p. 22 Dragon Images/Shutterstock.com.

Cataloging-in-Publication Data

Names: Miloszewski, Nathan.
Title: Plumbers and sewage workers / Nathan Miloszewski.
Description: New York : PowerKids Press, 2020. | Series: Getting the job done | Includes glossary and index.
Identifiers: ISBN 9781725300125 (pbk.) | ISBN 9781725300149 (library bound) | ISBN 9781725300132 (6pack)
Subjects: LCSH: Plumbing–Juvenile literature. | Plumbers–Juvenile literature. | Sewerage–Juvenile literature. | Sewage disposal–Juvenile literature. | Sewage-Purification–Juvenile literature.
Classification: LCC TH6124.M526 2020 | DDC 696'.1023–dc23

Manufactured in the United States of America

CPSIA Compliance Information: Batch #CSPK19. For Further Information contact Rosen Publishing, New York, New York at 1-800-237-9932.

CONTENTS

A WORLD OF PIPES

Pipes are everywhere! They run through our cities, buildings, and houses. That's a lot of pipes for plumbers and sewage workers to install, or put in place. We need pipes to carry water into and out of our houses so we can cook food and wash our dishes, brush our teeth and take showers, and wash our clothing.

You might not think about your pipes until there's a problem with them. Your tub drain might be clogged or maybe there's sewage backing up into your toilet. To fix problems with your pipes, a plumber or sewage worker will come to the rescue!

Your parents may be able to fix small problems with your pipes. However, larger problems mean they'll need to call a plumber.

>

BEING A PLUMBER

To be a plumber, you have to use your hands and your brain. Plumbers put in water pipes and natural gas lines. A plumber's job requires good math skills. They need to take measurements, do math problems involving fractions, and connect different sized pipes and fixtures. Plumbers use many different tools. Plumbing tools include pipe wrenches, copper or plastic piping, a welding torch, special glues, clamps, and testing equipment.

Plumbers can't just walk into a bathroom and tell you why your sink is clogged. They need to **investigate** first! Plumbers squeeze under the sink, into crawl spaces, and even underground to do their job.

Fascinating Career Facts

Plumbers also need to know how to read **blueprints** and understand local **building codes**.

Working in small spaces in all kinds of weather is a part of the job. Plumbers often work upside down, on their backs, and in small spaces.

A SMELLY AND DANGEROUS JOB

Sewage workers deal with the pipes that carry water out of houses and buildings. This water often carries human waste, which can make it dangerous. There are many miles of sewer pipes below the street and under your home. This is where sewage workers do their work. They keep our water clean and stop wastewater from backing up into roads or homes.

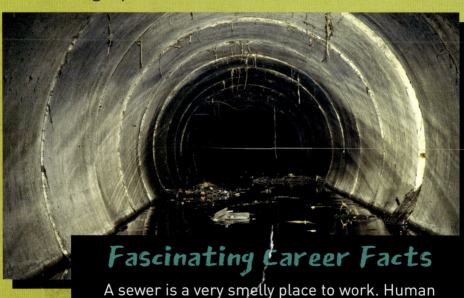

Fascinating Career Facts

A sewer is a very smelly place to work. Human waste, baby wipes, garbage, rotting food, and leaves are just a few things sewage workers find in our sewers.

One part of a sewage worker's job is to inspect sewers. Some sewers in older cities are hundreds of years old!

Sewage workers walk, **wade**, swim, and dive into streams of waste to remove sewage backups and repair any damaged sections of the sewer line. You may never see sewage workers. They open up manhole covers on the street and disappear into the dark tunnels below.

A BRIEF HISTORY

More than 2,000 years ago, the ancient Romans were the best plumbers in the world. They built **aqueducts** of stone to carry fresh water down from the mountains into the cities below.

Fascinating Career Facts

Plumbum means "lead" in Latin. This is where the word "plumbing" comes from. Lead was the metal used to make pipes many years ago. Lead is known to make people sick, so we no longer use lead pipes.

The Pont du Gard is a famous Roman aqueduct that crosses the Gardon River in southern France. It was built in the first century AD to supply the city of Nîmes with fresh water. In 1985, it was named a UNESCO World Heritage Site.

In the past 200 years, there have been many **innovations** in plumbing and sewage. These innovations help us stay clean and healthy and they are **affordable** for many people. However, in some parts of the world, people can't easily get fresh, clean water and don't have sewers to carry away wastewater. Without fresh, clean water, many people in these parts of the world get sick. Some people even die because of contaminated, or dirty and impure, water.

11

MAJOR MYTHS

One popular myth is that Sir Thomas Crapper invented the modern flushing toilet. Sir Thomas Crapper was a plumber who had some **patents** in the late 1800s. However, Sir John Harington actually invented the first flushing toilet in 1596.

Some people believe the myth that alligators live in sewers. However, sewers aren't good homes for alligators and they don't like living there. Alligators have gotten stuck in storm drains and large outdoor pipes while exploring. Don't be afraid, though, because they can't get into your house pipes. Those are too small for a big alligator to fit inside!

Fascinating Career Facts

Sir John Harington was a member of the royal court of Elizabeth I, a translator, author, and a wit, or person talented at making clever and amusing comments. Queen Elizabeth I was his godmother.

February 9 is "Alligator in the Sewer Day" to remember the discovery of an alligator in the sewer at 123rd Street in New York City in 1935.

13

A PLUMBER'S DAY

A plumber's day starts early in the morning and can run late into the evening if they take emergency calls. Plumbers may visit several homes or job sites in a single day.

One of the most common problems a plumber is called to fix is a clogged toilet. People flush everything you can imagine—from toys to toothbrushes—down their toilets.

Plumbers' jobs have a lot of **variety**. Different rooms have different types of sinks, drains, and pipes. Plumbers also spend time talking to local plumbing inspectors who make sure all of their work is up to code.

Fascinating Career Facts

Plumbing codes are a set of rules that say what a plumber can and can't do. They also make sure work is done well and helps your family stay safe and healthy. Plumbing codes are always changing.

14

A plumber may start their day working on a dishwasher, then work on a bathroom, and finish their day with a water heater.

IN THE CITY'S BOWELS

Sewage workers spend their days in the city's bowels, or inside parts. On a normal day, sewage workers perform routine plumbing **maintenance**. They put cameras into pipes to inspect the state of the pipes and to find common clogs such as cooking grease that's become hard, tree roots, and baby wipes. Sometimes even larger household items can be found.

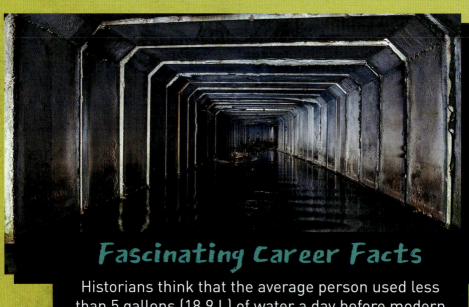

Fascinating Career Facts

Historians think that the average person used less than 5 gallons (18.9 L) of water a day before modern plumbing. Today, many people use about 100 gallons (378.5 L) a day—most of which winds up in the sewer.

Sewage workers use big trucks with hoses hooked up to fire hydrants to pump high-pressured water into sewers to keep them clean. Other times they have to suit up and dive through dirty water, animals, and smells to keep waste moving properly.

EDUCATION AND TRAINING

To become a plumber, you need a high school degree. Then you'll take plumbing classes at a trade school or technical or community college. From here, you can complete an **apprentice** program. Then you can get your plumber's license. Licensing is different in every state, but most licenses require you to have two to five years of experience and the passage of an exam.

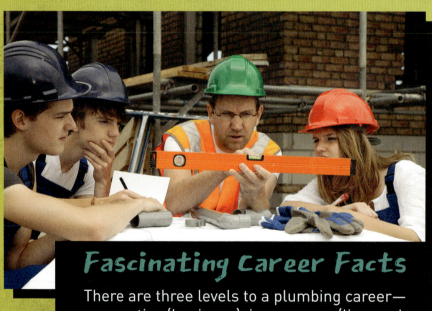

Fascinating Career Facts

There are three levels to a plumbing career—apprentice (beginner), journeyman (licensed and reliable), and master plumber (expert).

Becoming a plumber or sewage worker can be a mix of homework, tests, and learning from professionals at real job sites.

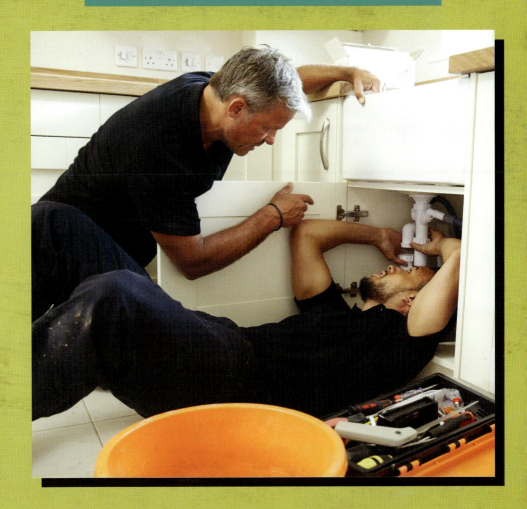

Sewage workers don't always need to complete a formal training program. If you want to work in a wastewater treatment plant you have to pass a test to get a **certificate** or become licensed in your state and then you'll get trained on the job.

HARD WORK PAYS OFF

Being a plumber or sewage worker is hard work. These jobs can be dirty, dangerous, and hard on your body. However, plumbers and sewage workers are paid well to do this work.

Beginner plumbers make around $40,000 per year and can make upwards of $100,000 when they're experienced. Sewage workers can make more than $60,000 per year.

A career as a plumber or sewage worker can lead to other job opportunities, too. One day, you might be able to train others, manage a team of workers, or even start your own plumbing company!

Plumbers and sewage workers often have the ability to choose which hours they work and who they work with. Some choose to work alone.

21

A BRIGHT FUTURE

Plumbers and sewage workers are some of the highest-paid construction workers. However, there's often more work that needs to be done than there are workers to do it.

Working in construction is perfect for people who don't want to go to a four-year college or university after high school. This allows workers to learn a trade and make more money sooner.

The number of jobs in these fields is expected to continue growing in the future, too. We have many pipes and problems will always arise. This means we'll always need plumbers and sewage workers!

GLOSSARY

affordable: Describing something that you have enough money to buy.

apprentice: A person who is learning from a skilled professional that has years of experience.

aqueduct: A man-made structure that moves water from one place to another.

blueprint: A mechanical drawing from an architect that shows detailed plans for the work that needs to be done.

building codes: Rules that set the standard for how work must be done to make sure it protects the health and safety of people who live and work in homes and buildings.

certificate: A piece of paper or document that proves that you did something such as finishing a class.

innovation: Something that is new, such as a new idea or a new way of doing something.

investigate: To try to find out the facts about something.

maintenance: To take care of something on a regular basis. Also, work done on something before it breaks.

patent: A document that protects an invention, or idea, from being stolen and used by another person or company without giving credit, or money, to the inventor.

variety: A mix of things to choose from.

wade: To step in or walk through something such as water or mud.

INDEX

WEBSITES

Due to the changing nature of Internet links, PowerKids Press has developed an online list of websites related to the subject of this book. This site is updated regularly. Please use this link to access the list: www.powerkidslinks.com/gtjd/plumbers